THE POETRY OF MENDELEVIUM

The Poetry of Mendelevium

Walter the Educator

Silent King Books

SILENT KING BOOKS

SKB

Copyright © 2024 by Walter the Educator

All rights reserved. No part of this book may be reproduced in any manner whatsoever without written permission except in the case of brief quotations embodied in critical articles and reviews.

First Printing, 2024

Disclaimer
This book is a literary work; poems are not about specific persons, locations, situations, and/or circumstances unless mentioned in a historical context. This book is for entertainment and informational purposes only. The author and publisher offer this information without warranties expressed or implied. No matter the grounds, neither the author nor the publisher will be accountable for any losses, injuries, or other damages caused by the reader's use of this book. The use of this book acknowledges an understanding and acceptance of this disclaimer.

"Earning a degree in chemistry changed my life!"
- Walter the Educator

dedicated to all the chemistry lovers, like myself, across the world

MENDELEVIUM

A fleeting waltz of subatomic charm,

MENDELEVIUM

In the heart of the periodic farm,

MENDELEVIUM

Mendelevium, a noble quest,

MENDELEVIUM

In the crucible of discovery's zest.

MENDELEVIUM

Born of stars in cosmic blaze,

MENDELEVIUM

In supernovae's fiery haze,

MENDELEVIUM

Synthesized in human hands,

MENDELEVIUM

Unveiling secrets of distant lands.

MENDELEVIUM

With atomic number ninety-five,

MENDELEVIUM

In the chart, it does arrive,

MENDELEVIUM

A rare creation, elusive and grand,

MENDELEVIUM

In laboratories, it makes its stand.

MENDELEVIUM

Named for Dmitri Mendeleev's grace,

MENDELEVIUM

Whose vision the elements embrace,

MENDELEVIUM

In his honor, its name proclaimed,

MENDELEVIUM

In the annals of science, forever framed.

MENDELEVIUM

A transuranic element, heavy and rare,

MENDELEVIUM

In the realm of the periodic fare,

MENDELEVIUM

Its isotopes, unstable and fleet,

MENDELEVIUM

Decay with time, their fate complete.

MENDELEVIUM

Yet in their fleeting existence, they reveal,

MENDELEVIUM

The essence of matter's cosmic zeal,

MENDELEVIUM

Unlocking doors to the unknown,

MENDELEVIUM

In the nucleus, a universe is shown.

MENDELEVIUM

Mendelevium, a pioneer bold,

MENDELEVIUM

In the quest for knowledge untold,

MENDELEVIUM

Pushing boundaries, expanding the mind,

MENDELEVIUM

In the endless pursuit of truths confined.

MENDELEVIUM

Its properties, a scientist's delight,

MENDELEVIUM

In experiments, they shine bright,

MENDELEVIUM

Revealing glimpses of fundamental laws,

MENDELEVIUM

In the labyrinth of nature's cause.

MENDELEVIUM

From Berkeley's labs to Dubna's halls,

MENDELEVIUM

In the pursuit of scientific thralls,

MENDELEVIUM

Mendelevium's saga unfolds,

MENDELEVIUM

In the quest for understanding's molds.

MENDELEVIUM

From its discovery to the present day,

MENDELEVIUM

In the tapestry of science, it holds sway,

MENDELEVIUM

A symbol of human curiosity,

MENDELEVIUM

In the quest for cosmic verity.

MENDELEVIUM

So let us raise our eyes to the sky,

MENDELEVIUM

Where stars above twinkle and fly,

MENDELEVIUM

In the dance of atoms, we find our home,

MENDELEVIUM

And Mendelevium's legacy, forever known.

MENDELEVIUM

ABOUT THE CREATOR

Walter the Educator is one of the pseudonyms for Walter Anderson. Formally educated in Chemistry, Business, and Education, he is an educator, an author, a diverse entrepreneur, and he is the son of a disabled war veteran. "Walter the Educator" shares his time between educating and creating. He holds interests and owns several creative projects that entertain, enlighten, enhance, and educate, hoping to inspire and motivate you.

> Follow, find new works, and stay up to date
> with Walter the Educator™
> at WaltertheEducator.com

www.ingramcontent.com/pod-product-compliance
Lightning Source LLC
LaVergne TN
LVHW012049070526
838201LV00082B/3867